What Is that Alligator Saying?

What Is that Alligator Saying?

A science book about the way animals talk to each other

by RUTH BELOV GROSS

Pictures by JOHN HAWKINSON

SCHOLASTIC BOOK SERVICES

NEW YORK • TORONTO • LONDON • AUCKLAND • SYDNEY • TOKYO

The author wishes to thank The New York Public Library for allowing her to use the Frederick Lewis Allen Memorial Room, where this book was written.

This book is sold subject to the condition that it shall not be resold, lent, or otherwise circulated in any binding or cover other than that in which it is published — unless prior written permission has been obtained from the publisher — and without a similar condition, including this condition, being imposed on the subsequent purchaser.

Text copyright © 1972 by Ruth Belov Gross. Illustrations copyright © 1972 by Scholastic Magazines, Inc. All rights reserved. Published by Scholastic Book Services, a division of Scholastic Magazines, Inc.

1st printing .... January 1972

Printed in the U.S.A.

*For
Willy
again*

People can tell each other things
in a lot of different ways.
Mostly they do it by talking.
People can also tell each other things
without talking at all.

They can use their hands.
They can make faces.
They can make funny sounds.
They can cry.

beep beep

Animals have ways
of telling each other things too.
Sometimes we say they are talking to each other.
But animals do not talk the way we do.
They do not use words.

A firefly is flashing its light.

Some animals do things
that other animals can see.

Some animals do things that other animals can feel.

Some animals make sounds
that other animals can hear.

A beaver is smacking its tail
on the water.

Some animals make smells that
other animals can smell.

A bear is scratching its back and leaving its
bear smell on the tree.

There is a word for
all the ways that people use
and all the ways that animals use
to tell things to each other.
The word is *communication*.

This book is about *animal communication*.

In this book, some sentences are written in capital letters, like this:
Give me some food.
The capital letters will show you what an animal might be "saying." Remember, though, that animals do not talk the way people do.

Many kinds of baby animals cannot take care of themselves. They need their parents to feed them and to keep them safe.

These baby animals have special ways to tell their parents when they are hungry or in trouble. And the parents have special ways to warn the babies when there is danger.

Baby alligators communicate with their mothers by grunting. Whenever a mother alligator hears her babies grunting — *umph, umph, umph* — she comes to them right away.

The first time a mother alligator hears her babies, she cannot see them. Their little grunts are coming from a pile of mud and old leaves. That is where the mother alligator laid her eggs many weeks ago.

Now the mother alligator goes to the pile of leaves. She digs it open with her long alligator snout. And there, under the leaves, are her babies! She helps them get out of their muddy nest.

If the baby alligators couldn't communicate with their mother, what would happen? Maybe they would not be able to get out of their muddy nest by themselves.

Mother alligators also communicate with their babies by grunting. *Umph, umph, umph* — a mother alligator is warning her babies of danger. The babies hide in the water when they hear this sound.

If a mother alligator couldn't warn her babies of danger, what would happen to them?

A mother hen is with her chicks almost all the time. She keeps them warm under her feathers. She helps them find food. And she keeps them safe from other animals.

What makes the baby chicks and the mother hen stay together? They stay together because they "talk" to each other.

When the mother hen goes *cluck-cluck-cluck*, the chicks come running to her. When the chicks go *peep-peep-peep*, the mother hen comes to them.

Scientists did not always know what made a mother hen come to her chicks. Did she come to them because she saw them — or because she heard them?

One day a scientist made an experiment. He put a chick under a big glass bowl. The mother hen could see her chick but she could not hear it. No one outside the bowl could hear it.

Peep-peep-peep went the chick, but the mother hen walked right past the bowl.

Then the scientist put the chick behind a wooden fence. The mother hen could not see her chick behind the fence. But she could hear it.

Peep-peep-peep went the chick. The mother hen heard it and went straight to the fence.

The experiment showed the scientist that the mother hen came to her chicks because she heard them.

Animals cannot say "Watch out!" the way we can. They have their own ways of warning each other that danger may be near.

A beaver uses its tail to say "Watch out!" It lifts its tail up over its back. Then it smacks its tail *hard* on the water of the beaver pond.

The sound can be heard far away. It tells other beavers that an enemy is near.

When the other beavers hear the smacking sound, they dive into the pond. But first they smack *their* tails on the water.

Smack! Whack! Smack! The beavers are passing the danger signal along.

Crows are noisy birds. They go *caw caw caw* a lot, and they make other noises too.

Hubert Frings and his wife Mable Frings are two scientists who studied "crow talk." They wanted to find out how crows tell each other

<small>Watch out — danger!
Fly away quickly!</small>

So they hid microphones near a bunch of crows, and they attached the microphones to a tape-recording machine.

After a while Dr. and Mrs. Frings had a lot of different crow sounds on their tapes. Then they played the sounds back to the crows over a loudspeaker. They played the sounds one at a time.

They played the first sound back to the crows. Nothing happened. They played another sound, and again nothing happened.

Then they tried a third sound. When they played this sound, crows came flying from all over.

They tried one more crow sound — and this time all the crows flew away! The crows flew away every time that sound was played.

Now Dr. and Mrs. Frings knew which sound meant "watch out — fly away." They named this the "alarm call."

And they also knew which sound made the crows come together. They named this the "assembly call."

An ant has two feelers sticking out of the top of its head. It feels things with its feelers, and it smells things with them too.

Ants use their feelers when they communicate with each other.

When two ants meet, they stop and tap each other with their feelers. They are saying

Y<small>OU SMELL ME</small>
<small>AND</small> I <small>WILL SMELL YOU.</small>

If they are both from the same nest, both ants will have the same smell. If they are from different nests, the two ants will have different smells. Then they may fight with each other.

An ant will often stroke the face of another ant with its feelers and legs. That is the way an ant says

          GIVE ME SOME OF YOUR FOOD.

Then the other ant will give it a drop of food. This is extra food that the ant carries in a special stomach. Many ants carry this extra food. They share it by passing it from mouth to mouth.

Sometimes an ant finds a piece of food that is too big to carry. The ant gets excited and runs back to its nest.

On the way, it stops many times. It presses its body against the ground and leaves a little spot of smelly stuff. The spots make a smelly trail on the ground.

When the ant gets back to its nest, the other ants rush to the smelly trail. They follow it with their feelers. The trail tells them

THIS WAY TO THE FOOD.

Other animals also have ways of telling each other

THERE'S FOOD HERE — COME AND GET IT!

Flies leave a special smell on the food they visit. The smell helps other flies find the food.

Here is an experiment someone tried. First he found a place where there were some flies. He put some sugar where the flies could get to it. Then he waited.

After a long while, one fly found the sugar. But soon after that, many flies came.

Why did the other flies come? Did they smell the sugar? No — sugar does not have a smell.

They came because they smelled the fly smell on the sugar. Every fly that came to the sugar left some fly smell there.

Sometimes gulls make a special kind of sound when they find food. This sound brings other gulls over in a hurry. Scientists call it the "food call."

A gull gives the food call when he finds plenty of food. If he finds only a little bit, he doesn't make a sound. He just eats the food himself.

There is a time in the life of almost every animal when it gets together with another animal and mates with it.

A dog finds another dog to mate with, a beaver finds another beaver, and a firefly finds another firefly. Every animal finds an animal of its own kind to mate with.

If animals couldn't find mates, what would happen? There wouldn't be many animals in the world.

Animals find their mates by communicating. This is how crickets, moths, fireflies, and birds communicate to find a mate.

Male crickets sing to female crickets. They sing by rubbing their front wings together. They sit in meadows and forests and sing.

Each kind of male cricket has its own kind of song. Each song says

COME MATE WITH ME.

If a female cricket is ready to mate, she will come to the male cricket when she hears him calling.

Moths find each other in a different way.
When it is time for moths to mate, the female moth wears a special smell. The smell makes the male moth come to her.

People can't smell this moth-y smell. Female moths can't smell it either. But male moths have no trouble smelling it.

A male moth will fly to a female moth even if she is miles away.

On warm summer nights in grassy places, fireflies call their mates by flashing little lights on their bodies.

First a male firefly flashes his light. He flashes it on and off, on and off. He is saying Here I am. I am looking for a female firefly to be my mate.

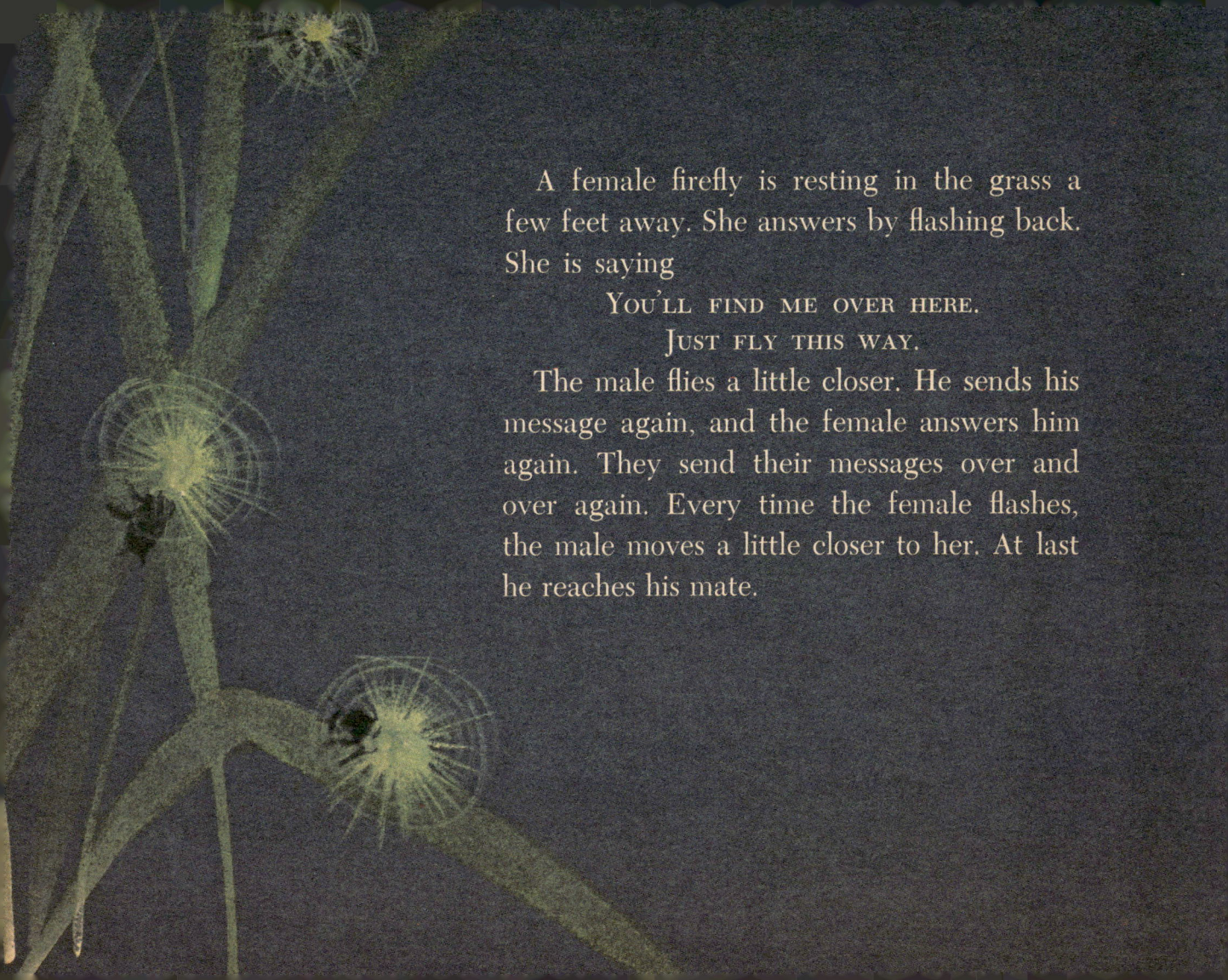

A female firefly is resting in the grass a few feet away. She answers by flashing back. She is saying

>You'll find me over here.
>Just fly this way.

The male flies a little closer. He sends his message again, and the female answers him again. They send their messages over and over again. Every time the female flashes, the male moves a little closer to her. At last he reaches his mate.

Springtime is the time of year when the birds sing loudest.

People used to think the birds were saying "Spring is here! Spring is here!" But the birds aren't saying anything like that. It is time for them to mate, and the males are making important announcements.

Male robins are making announcements to other robins. Male sparrows are making announcements to other sparrows. And male bluebirds are saying things to other bluebirds.

Each male is telling the other males of his kind
>THIS PLACE IS MINE! DON'T COME TOO CLOSE OR I WILL FIGHT!

And he is telling the females
>I AM READY TO MATE. YOU CAN FIND ME HERE.

A man named Eliot Howard figured this out. Every morning before he went to work, Mr. Howard used to go outdoors to watch the birds near his house. He kept a notebook and wrote down everything he saw and heard.

Mr. Howard saw the same thing year after year. In the spring, the male birds left their flocks. Each male picked a place for himself and started to sing. He sang very loud.

If another male of the same kind came near, he was chased away. But if a female came near, she was allowed to stay. Then the two birds stayed together to mate.

By watching and listening, Mr. Howard learned that the loud singing kept the other male birds away and brought the females closer.

The place that a bird picks for his very own is called his "territory." His singing tells other birds that the territory belongs to him.

Other kinds of animals have territories too.

Each kind of animal has its own way of saying "This is my territory."

A male fiddler crab has one regular-size claw and one big claw.

Some kinds of male fiddler crabs say "This is my territory" by waving the big claw. The male crab sits near his crab hole and waves. He waves only at special times, when he is ready to mate.

If another male comes along, the crab may wave even harder. Then the other crab will go away — or he may come closer. If he comes closer, there may be a fight. The fight usually looks worse than it really is.

If a female crab sees the male waving his big claw, and if she is ready to mate, she may come over to him. Then she follows him down into the crab hole.

The waving doesn't seem to mean "Keep away" to her.

A brown bear leaves a smell around the edges of his territory. This tells other brown bears

You are about to enter my territory.

The smell usually makes a male bear go away. But if it is mating time, a female bear may decide to hang around.

Where does the bear leave his smell? On tree trunks. He stands on his hind legs and rubs his back on a tree trunk.

There is enough bear smell in the bear's hairy back to do a good job.

A note from the author

This book tells you some things about animal communication, but it does not tell you everything. If you want to, you can find out more about animal communication.

You can find out some things by reading.

Maybe you will want to read about bees. Whole books have been written about bees — how they do little dances to tell other bees where to find food.

Maybe you will want to read about dolphins. They communicate with each other by making whistling and clicking sounds under the water.

You can also find out more about animal communication by watching what animals do. You can watch spiders and flies and ants, and you can watch dogs and cats and birds.

Maybe you will find out something that nobody knew before.

Index

Index of animals

Alligators, 12-14
Ants, 8, 24-26
Bears
   Brown bears, 9, 40-41
Beavers, 9, 18-19
Bees, 42
Birds
   Chickens, 15-17
   Crows, 20-23
   Gulls, 28
   Songbirds, 34-37
Brown bears, 9, 40-41

Chickens, 15-17
Crabs
   Fiddler crabs, 38-39
Crickets, 30
Crows, 20-23
Dolphins, 42-43
Fiddler crabs, 38-39
Fireflies, 8, 32-33
Flies, 27
Gulls, 28
Moths, 31
People, 7

Other subjects you may want to look up in this book

Baby animals and their mothers, 11-17
Alligators, 12-14
Chickens, 15-17

How animals communicate
Introduction, 7-10
Sight
   Fiddler crabs, 38
   Fireflies, 8, 32-33
Smell
   Ants, 24, 26
   Brown bears, 9, 40
   Flies, 27
   Moths, 31
Sound
   Alligators, 12-14

Beavers, 9, 18
Birds
   Chickens, 15-17
   Crows, 20-23
   Gulls, 28
   Songbirds, 34-37
Crickets, 30
Dolphins, 42
Touch
   Ants, 8, 24-25

Experiments
Chickens, 15-17
Crows, 20-23
Flies, 27

What do animals communicate about?
Danger, 14, 18-23
    Alligators, 14
    Beavers, 18
    Crows, 20-23
Food, 25-28, 42
    Ants, 25-26
    Bees, 42
    Flies, 27
    Gulls, 28
Mating, 29-36
    Birds, 34-36
    Crickets, 30
    Fireflies, 32-33
    Moths, 31
Territory, 35-40
    Birds, 35-37
    Brown bears, 40
    Fiddler crabs, 38